Carnac, Locmariaquer and Gavrinis

TEXT
CHARLES-TANGUY LE ROUX
General Heritage Conservator

PHOTOGRAPHS
YVON BOELLE

Above.
Entry to the Gulf of Morbihan, between Port-Navalo and Locmariaquer; foreground, the Petit-Mont cairn dominates the Crouesty cove.

Cover.
he Carnac alignments, a detail of Gavrinis, entrance to the Table-des-Marchands, the Mané-er-Hroeg stele.

Fourth cover page
The Gavrinis cairn.

Contents

TRANSLATION 5/5

ÉDITIONS OUEST-FRANCE

The Neolithic, a new age

New behaviour in a completely changed world

The Neolithic, the period when the whole of Western Europe became covered with megalithic monuments, is much more than the "age of polished stone" of the first commentators on pre-history. It corresponds to a new way of living for humanity, which ceased simply to take from nature its means of subsistence by hunting, fishing and gathering and started to organise production with agriculture and stock-breeding. Other technical innovations, however, accompanied the change in attitude: ceramics, weaving, basketwork and the polishing of hard stones. All this implied at least a partially sedentary habitat, leading to architecture and the territorial concept. At the same time elaborate social structures emerged and structured religious thought developed, already perhaps bearing the seeds of today's major religions.

Locmariaquer, the group ▼ formed by the Grand-Menhir, the circular cairn of the Table-des-Marchands and the long Er-Grah tumulus extend not far from the river Auray along which the large slabs necessary for their construction probably travelled.

Imagining the world in which the first persons made their settlements is not easy. It is known that after the grip of the last ice age, towards 18000 to 6000 BC, the climate began to become warmer, at first weakly and haphazardly, then rapidly from about -8000. This was the period when the Neolithic process began in the Near East. Subsequently an "optimum climate" was reached, a little warmer even than our times, between –5000 and –3000 – corresponding to Neolithic settlement and extension into Western Europe. The drastic climatic change had two major effects: the rise in sea level and the forest's re-conquest of the land.

At the beginning of the Neolithic age, at the foot of the present cliffs broad coastal plains, fertile and attractive for the first farmers, must have remained but they were obviously "at the mercy of the sea". It is conceivable that dwellings grew alongside the cultivated land but the "monuments to eternity" like the megalithic stones were

▼ *The Gulf of Morbihan with its islets "as numerous as the days of the year" shows how the sea covered a formerly fertile basin crossed by the estuaries of the present Vannes and Auray streams.*
Photo: Yannick Legal.

NEOLITHIC ROOTS

The adventure began in the Near East starting in the eighth millennium BC; in this region in particular the ancestors of our cereals grew – notably wild wheat and barley – and various pulses (peas and related plants), while at the same time herds of wild goats and sheep flourished. With the addition of some vegetables, beef and pork (domesticated from the auroch and the wild boar), the foundations were already laid for what was to become, almost to our days, the basis for European subsistence agriculture.
From the Near East, the new way of life reached Turkey and then the Balkans. From there it spread throughout the Danube basin and extended along the Mediterranean coast. Towards the middle of the sixth millennium BC the first of these major movements reached Alsace and the second penetrated into Provence and Languedoc. Traces still in existence reveal a possible "pioneer" Neolithic tendency along the Morbihan coast from –5000, even a little earlier, the decisive steps being taken in the following centuries.

The two modest cups ▲▶ with curiously misshapen opening (Vannes museum) probably feature among the pots of Morbihan's first farmers.

more often erected beyond the reach of the potential menace.

Mesolithic hunters lived in balance with the forest environment, which they exploited for its resources in game and vegetation. On the other hand, for Neolithic farmers, the forest was an environment to conquer. It sheltered alarming wild beasts and also rather different men – probably not always benevolent. Also, however, it represented land to clear and to provide wood, the all-purpose material (from fire to frames for houses and scaffolding for megalithic work-sites). Using axes of polished stone, the Neolithic man was able to start to "devour the forest".

The first farmers of Armorica

Producing food is much more exacting than simply harvesting from nature; it also ensures better protection against hazards – climatic or otherwise – and above all enables a much larger population to be fed from the same land. It is estimated that in general the average Neolithic population density was one or two inhabitants per square kilometre in temperate Western Europe; this amounts to

30,000 to 50,000 people for Brittany, with major disparities between regions already highly populated (such as the Morbihan coast in all probability) and others almost deserted in the interior.

The first farmers mastered the art of ceramics, as borne witness by the few potteries from the epoch which have been found, for example near Arzon. Their beasts included sheep and goats (the progeny of breeds already domesticated in the Near East, these animals being unknown in west European wild fauna), and also cows and pigs (which, in their case, it was frequently possible to cross with local auroch and wild boar breeds). Similarly their cereals (wheat, barley) derived from seeds originating in the Levant. Quite quickly, however, empirical selection techniques distinguished these plants from their wild ancestors, while at the same time new crops appeared (for example – already – buckwheat).

There is every reason to think that the land taken from the forest became exhausted in a few years, then being abandoned to moorland and copses which after several decades, sufficiently regenerated the soil for a new clearing, often by burning. At this pace, it is conceivable that the forest diminished rapidly (this can

be seen for example in the traces of pollen in the peat bogs and the growth rings of pieces of charcoal found in fireplaces).

The first farmers, with still rudimentary tools and techniques, preferred light soils, easy to work. Unfortunately the latter were to be found in the coastal plains now under water, or the hills in the interior which have since been eroded, so that scarcely any traces of their dwellings are to found.

THE SEA AND THE FOREST

In the glacial period, enormous stocks of water were "frozen" (both in the real and figurative meaning of the term) on the continents and the level of the oceans fell. At the height of the last ice age about twenty millennia ago, the Channel was dry and the Atlantic coast extended about 100 kilometres beyond the present Morbihan coast. With the thaw, the glaciers melted and the water returned to the oceans whose level rose, slowly at first then more rapidly (up to more than 1 metre/century!), reaching –10 metres around –5000, i.e. at the dawn of the Neolithic age in our regions. The movement subsequently slowed down; it became more irregular and the level became more or less stable at the beginning of our era.
When the last ice age was at its strongest, our regions' vegetation would have resembled the present tundra, without trees except in rare, extremely sheltered places. With the thaw, a "pioneer" forest with pines and birches managed to seed itself from these refuges, soon to be followed by "mixed groves" combining our main deciduous species. The primary forest (only a few intact remains in Central Europe give an idea of what it was like) soon spread everywhere, only sparing the mountains, flooded valleys and the most exposed coasts.

▼ ***The form of some of the Carnac menhirs reveals their removal from their natural outcrops in the neighbourhood; they were already shaped by geological erosion.***

Thirst for the monumental

The little that is known of Neolithic houses shows that at least some of them already far exceeded the strict necessity for providing shelter. They simultaneously reveal a taste for the monumental, advanced social organisation and an assertion of identity. But, far better than wood, making use of colossal stones responded to additional concerns of a metaphysical nature, both by the spectacular aspect which they confer on the constructions and by the impression of durability – even eternity – which they convey. Perhaps the symbolic value of "stony matter" can be seen in them, even if these huge monoliths do not constitute the entirety of the monumental structures and not all of them were visible in the completed building. It is hardly surprising therefore that these megalithic monuments are to be found almost everywhere in Western Europe, from the south of Scandinavia to the south of Spain and along the shores of the Atlantic to the foothills of the Alps, from the plains of north-east Germany to the islands of the Western Mediterranean.

The classic names of menhirs and dolmens remain useful for making a distinction between the two major families of megalithic monuments, those with standing stones (isolated or in more or less complex groups – in circles or aligned) and those of the megalithic tombs – themselves very diversified, as in the conception of the burial chambers properly speaking (closed caverns or rooms remaining accessible) and in the mausolea protecting and enhancing them (cairns or mounds).

In this sector of Kermario, ▼ the general organisation of the lines suggests a cultural positioning, but sometimes (here on the extreme left), the "out of alignment" are perhaps the relics of a different former positioning.

When one compares all these achievements, the investments they represent in time, energy and motivation and the archaeological remains they have revealed, one realises that their meaning is probably very diverse and that the funerary function usually ascribed to them – generally correctly – was very often only one of their aspects.

Recently decades' excavations have shown that the megalithic tombs – in Armorica as elsewhere – were built according to the rules of a simple and empirical, but perfectly mastered, know-how. The mass of cairns and mounds was structured and their surface dressed with facing to balance the internal pressure, whereas their interior crypts were made from stone slabs carefully assembled or completed with dry masonry. The finest technical achievements are certainly the encorbelled "false copula" coverings for some of the chambers; a few of these, notably in Finistère, have braved the millennia to reach our times intact.

▲ *Facing Locmariaquer, the Arzon peninsular also bears a large number of monuments. Although spoiled by a blockhouse, the Petit-Mont cairn is probably the most spectacular among them.*

Similarly, excavations of menhir sites have shown the care with which stones were balanced before a wedge – often astonishingly light – fixed them into place.

In many cases the megaliths were constructed with material available on the spot, in general taken from outcrops rather than quarried. But their obviously having been transported, and this sometimes defies understanding as at Locmariaquer – clearly shows that the site chosen was not simply the result of geological expediency. Here again, handling techniques – simple but operated with great care – have been reconstructed from observations made during the excavations; they imply the existence of experienced specialists who directed the work.

NEOLITHIC DWELLINGS

In all of Brittany, the only ancient Neolithic house, where enough of its components have been found to make a reliable reconstruction, is on display near Saint-Etienne-en-Coglés, to the north-east of Rennes. The building, with a framework of wooden stakes driven deeply into the earth, was trapezoidal in shape; it reached 22 metres long by 10 metres wide at its main façade (facing east), but only 4 metres at its western gable, facing the prevailing wind. This type of house, thatch covered with roofing also sloping towards the west, was widespread at the time throughout the Paris basin. Its dimensions indicate that the basic group which it protected was an extended family of several couples (probably related) accompanied, at least in the harsh season, by animals kept for reproduction.

More recent by a millennium and a half are the "big houses" at Pléchâtel, to the south of Rennes; the highest exceeded 100 metres in length by 12 metres in width and its roof would have reached 10 metres in height, some weight-bearing pillars approaching 1 metre in diameter. It is thought that its corresponding social structure was even more extensive, embracing numerous family units under its single gigantic roof.

The countryside of Carnac

SERPENT WORSHIPPERS AND CELTISTS

At the end of the eighteenth century, absorbed by the Enlightenment but also replete with symbolism and the esoteric, there arrived from England the ideas of the "cult of the serpent", "temple of the dragon", the "celestial theme" and the work of the "Celts our forebears" attached to the megaliths. Though these ideas were a relatively passing fashion, the Celtic Revival, directly competing with romanticism, met with widespread and lasting success; the outburst put a brake on objective research into the megaliths for two-thirds of a century.

The land and its stones

The huge municipality of Carnac spreads along the shore of Quiberon Bay between its neighbours Plouharnel and La Trinité-sur-Mer. The sub-soil, composed of clear and slightly foliated granite, is studded with tabular blocks suitable for the building of megalithic monuments, particularly on a number of small hillocks, which often bear the "Mané" toponym in Breton, included in the name of a number of the monuments.

Strange to relate, the extraordinary concentration of megalithic monuments in the Carnac countryside seems hardly to have attracted the attention of the learned before the second quarter of the eighteenth century. Popular tradition mainly refers to the local patron saint, Cornély, who turned a pursuing army of Roman legionaries into stone. Venerated as a protector of cattle throughout the south of Brittany, Saint Cornély probably corresponds to the Pope Cornelius, persecuted for having the denounced animal sacrifice in Rome in the third century.

From 1750 onwards, however, descriptions and theories multiplied. Already in 1764, the Count of Caylus expressed the idea of the extremely ancient provenance of the megaliths but others, carried away by their enthusiasm, were not always so perceptive.

▼ ***The "sky effects" on the alignments make it easy to understand how a whole group of legends have gathered round these stones.***

All these speculations did not prevent researchers from going into the field. The first "master plan" of the region's monuments was drawn by Murray and Vicars in 1832. Subsequently the major tumuli were explored by the polymath Morbihan society in the years 1850-60. In the last third of the century, William Lukis with Henry Dryden drew a number of plans of the monuments – still untouched by any restoration – whereas James Miln undertook several excavations; their booty started the Carnac museum.

▲ *According to attitude and point of view, the alignments sometimes give the impression of immutable order, sometimes a milling mass of stones that resemble living creatures.*

A stone regiment

Menhirs do not only exist in Carnac, far from it; thousands of stones, having survived the endless causes of destruction awaiting them, still stand elsewhere in Brittany and in many other regions. What makes Carnac exceptional is the unrivalled extent which the phenomenon takes with the broad "fields of menhirs", which one finds still spectacularly repeated in neighbouring communes. All the guides describe the three large Carnac alignments – the Menec, Kermario and Kerlescan – which lengthened by the Petit-Menec of

TWO PRESTIGIOUS MUSEUMS

In 1881 James Miln, a Scottish archaeologist, bequeathed to the Carnac municipality the objects collected during the excavations which he had carried out in the region. Conservation and additions to the collection were the work of Zacharie Le Rouzic, his young pupil from Carnac, up to his death in 1939. In 1985 the museum was installed in the former presbytery of Carnac, completely rearranged for the purpose. Human presence in the Carnac countryside, from Palaeolithic times up to the Roman era, is presented in a modern and instructional way but the Neolithic is undoubtedly the star with its display of a large number of megalithic tools, a reproduction of the daily life of the time and a spectacular gallery of crafts from ornamented megalithic slabs.
Installed in the "jolly chateau" at the heart of the old town, the Vannes pre-history museum bears witness to the activity of the polymath Morbihan society, one of the oldest learned societies in France, founded in 1826 and still active.
In particular it presents prestigious displays of the large tumuli explored at the end of the last century, but it also contains many other riches, in particular an impressive natural history collection.

La Trinité-sur-Mer, over almost 4 km form a nearly continuous barrier from the south-west to the north-east between the shoreline and the hinterland. Further north, there is also the group from Kerzerho to Erdeven, the remains still extending over two kilometres, and also the smaller series in Plouharnel and Saint-Pierre-Quiberon. Comparisons clearly show that they have coherent organisation with an overall meaning, even if each one of the stones which they contain possesses its own value.

The classic positioning of the Carnac groups comprises more or less parallel lines (up to about ten) and variable length (from about one hundred metres to one kilometre at the most), in which the menhirs succeed one another regularly. The lines in general start from a small height, still sometimes occupied by a ring (ovoid or subquadrangular), bordered by standing stones almost joined together. The blocks used remain rough, even when they are very irregular, and the traces of erosion they reveal show that they have been taken from outcrops rather than quarried. The largest are grouped at the head of the lines where some of them can exceed three metres in height, the average size then gradually decreasing.

Two frequent common elements in primitive religious sanctuaries are thereby combined: a sacred area, clearly defined, and a sacred pathway enabling it to be reached

◄ ***The aerial views of Kermario (above) and Menec (below) clearly show the alignments' general organisation, orderly but far from being as precise as sometimes claimed.***

▲ *The Kerlescan alignment at Carnac.*

◄ *The Kerzerho alignment at Erdeven.*

▲ *The "Manio Giant", Carnac's largest menhir which seems to make the link between the Kermario and Kerlescan alignments.*

in a straight line. Lines and rings can also exist separately: at Saint-Pierre-Quiberon they are close but independent, whereas the Crucuno ring does not seem to have any accompanying lines. Lastly, the way in which the stones overshadow one another from one line to another or reveal transversal perspectives, when one walks through the group, is perhaps neither fortuitous nor meaningless.

It is clear that "Carnac was not made in a day". Like any major architectural work, a field of menhirs is the result of a complex process of adding – subtracting – transforming: here and there, some menhirs stand out by comparison with their neighbours, owing to

A small alignment ▲ in the Carnac neighbourhood: the Petit-Moulin at Plouharnel.

One of the two Kerderff ▼ menhirs, to the west of Menec.

their size (the Menec "giant") or their position (at Kermario); they are perhaps the remains of previous or transitory states. In the present state of knowledge, a single chronological fact is irrefutable: the eastern part of the Kermario alignments blithely crosses the Manio burial mound, which suggests that the latter had already lost a large part of its meaning when the menhirs were planted over it. But if it was already an ancient monument (it dates back to the middle of fifth millennium BC) that hardly narrows the top end of the possible bracket. On the other hand, the sighting near Kerlescan of a tomb with lateral entry seems to indicate that the area of the alignment was still attractive in the second half of the third millennium, even if the "golden age of the megalith" had probably by then already come to an end.

Isolated menhirs and small alignments

The large groups of Menec, Kermario, Kerlescan and Kezerho must not make us forget the smaller groups like the alignments of Saint-Pierre-Quiberon, Sainte-Barbe, Moulin at Plouharnel and Keriaval in Carnac, apparently isolated and scattered menhirs. Other stones appear as satellites of the large groups: north-west of Menec, an isolated block is apparently placed in "close file" a few dozen metres from the western part of the alignment, another seems to be "watching the situation" a short distance from the edge of the Criffol mound. Two very beautiful monoliths stand in a clearing in Kerderff a little more to the west. Between Kermario and Kerlescan, the "Manio giant", the region's largest monolith, nearly 6 metres high, seems to make the link between the two large alignments, whereas to the north of Kerlescan about forty small menhirs draw a broad curve opening to the east. Similarly, to the north of the main Kerzerho field at Erdeven, an accessory line of about twenty blocks stretches northwards starting from several giant monoliths.

Linked burial grounds and dolmens

It is generally believed that almost everywhere in Brittany the tombs were linked to the megalithic alignments, but it is often difficult to say whether it was the menhirs which crossed their location or if, on the other hand, the presence of mausolea inspired the erection of stones in their neighbourhood; perhaps, indeed, the two proceedings took it in turns.

The tumular burial grounds, which, with their mass, seal the tomb or tombs they cover, are the most frequently linked. In Carnac, the best known is the one at Manio, bound by alignments, as we have seen. Surrounded by a small wall, it covered two graves and encased the base of a fine menhir. Protected by the monument's earth, it has preserved five magnificent engravings in serpent form, which seem to correspond to a similar number of polished axes planted in the ground at its feet. Far from being isolated, the mound was part of a veritable necropolis spread among a half-dozen monuments including one, which, now reduced to its surrounding stones, is none other than the famous "quadrilateral" situated a few paces from the large Manio menhir. A little further to the east a large mound, about sixty metres long, closes the north edge of the Kerlescan circle. At Erdeven, the Lannec-er-Gadouer mound at

▼ *Close to the "Giant", the Manio "quadrilateral" is in fact the surroundings of a flattened tumular mound.*

Surmounted by its chapel, ▲ the Saint-Michel tumulus is probably the most colossal tomb in Armorica; its prestigious funerary furnishing is now in the Vannes museum.
Photo: Charles-Tanguy Le Roux.

the extreme eastern end of the Kerzerho lines has recently been excavated; like the one at Manio, it has been proved to be very early in the region's megalithic sequence, probably dating back to the middle of the fifth millennium BC.

The most striking combination, an alignment with a tomb of the dolmenic type (i.e. provided with an access to the funerary room), is to be found at the head of the Kermario alignment where a dolmen with corridor, still surrounded by the remainder of its protecting cairn, stands at about thirty metres from the first menhirs, most likely formerly being part of a vanished ring. Comparable combinations, looser but nonetheless significant, are also to be seen at Keriaval and at Mané-bras in Erdeven.

Dwellings of immortality

125 metres in length, 60 metres broad and 10 metres high, the Saint-Michel tumulus forms a veritable artificial hill which, with the chapel which surmounts it, dominates the whole countryside of Carnac. Its exploration started in 1862 and finished between 1900 and 1907 but, limited to a few shafts and galleries, today shut for safety reasons, it remains only partial. It is, however, known that this huge monument was made with a mass of clay compressed between an axial core of stone and a facing of stones (today remodelled to form the terrace supporting the chapel). In the centre of the monument, a huge cavern is surrounded by other more modest crypts, whereas a small corridor dolmen occupies the eastern extremity. The central tomb contained prodigious furnishing now preserved in the Vannes museum: 11 large axe blades in polished pyroxenite and 25 smaller ones in clear fibrolite, 97 pearls and 10 seashell pendulums.

One is astonished by such an accumulation of prestigious objects, probably of high symbolic value but in proportion to the effort devoted to constructing a mausoleum intended to arouse man's admiration and envy, while perpetuating the memory of their possessor, a personage of exceptional quality. Moreover the Saint-Michel tumulus is not isolated. A half-dozen comparable giant tumuli gather between Arzon ("Caesar's Mound" at Tumiac), Locmariaquer and Carnac (where, at Moustoir, there is a second monument of this type). Like the small burial mounds, these giants probably date back to the middle of the fifth millennium BC.

Tombs for the living

The dolmenic tombs reveal a completely different funerary process, since they contain a means of reaching the funerary chamber through their cairn or protecting mound; devotional visits are therefore possible, even the laying of additional bodies. In the most ancient of these tombs, which in the case of Morbihan probably dates back to the end of the fifth millennium BC, the chamber is simple, circular or polygonal. The two best examples in Carnac are provided by the Kercado dolmen (where the protecting cairn has kept the main part of its original volume) and by the group of three dolmens at Mané-Kerioned.

In the more evolved relics, the chamber subdivides or is enclosed, which suggests more elaborate funerary rites, needing specialised areas; this layout is in particular to be found at Keriaval or, in the commune of Erdeven, at Mané-Groh and Mané-bras. The fine Rondossec group at Plouharnel bears witness to an intermediary situation since, on the three dolmens placed side by side in the same cairn, only one possesses a small cell adjacent to the main chamber.

Lastly, the final Neolithic evolution, which excessively elongated the chamber to the detriment of the other structures, is now only represented at Carnac by a tomb with lateral entry, situated to the northeast of the Kerlescan alignments. A different example is to be found at Locmariaquer, with the outstanding Pierres-Plates tomb.

▼ *In one of the Mané-Kerioned dolmens, the light which filters through the corridor reveals several ornamented slabs.*

▼▼ *At Keriaval, the burial space opens into small independent chambers bearing witness to the development of beliefs and funerary practices.*

The Locmariaquer peninsular

On a tongue of land, which juts out to close the western part of the Morbihan gulf, a dozen important megalithic monuments are gathered. As at Carnac the sub-soil is mainly fine grain granite but in this case it was not to the builders' liking; they did not hesitate to bring in the finest blocks available in the region, even having to transport them for kilometres.

Why "there and not elsewhere"?

A collection of this kind, needing the repeated exercise of technical skill, shows an extraordinarily strong will to assemble in this place such grandiose examples of faith and human pride. People have for a long time been lost in conjecture on the reasons for this choice and theories – both serious and absurd – have not been lacking. Research in recent years, however, is beginning to raise some corners of the veil.

The formation of the present sea-bed indicates that at the beginning of the Neolithic age, when the sea was still several meters below today's level, the Mor-bras must have formed a huge well-sheltered gulf, open to the south-east and bordered by fertile sandstone plains, whereas the present Morbihan was only interspersed by the estuaries of the Vannes and Auray streams, enabling nonetheless access to the hinterland by water.

At Kerpinher point, at the south of the peninsular, a study of a "sea-

In front of the Locmariaquer ▼ peninsular, the rocky expanses which become uncovered at low-tide must, in the Neolithic Age, have been covered with fertile alluvial soil.

The four fragments of the Grand-Menhir ► lying at the head of a series of interlocking stones reveal an alignment destroyed since Neolithic times.

At the base of main fragment ►► the part intended to be planted in the earth has remained untouched except for a small socket carefully hewn at its end.

Photo: Charles-Tanguy Le Roux.

bordered peat-bog", now submerged, reveals that at the end of the sixth millennium BC the region had been vigorously cleared and cultivated; it is now recognised as the most ancient example of Neolithic penetration into Brittany.

Almost in the same era under the site of the present Er-Grah tumulus, an obviously sacrificial ditch held the carcasses of two oxen. Already domesticated but still very close to the wild auroch, these animals probably represent the start of stock breeding in the region. Their burial indicates that the tip of the peninsular was already a place devoted to a religious cult.

It is therefore possible to imagine that the first Neolithic colonists, coming for example from the Loire estuary, disembarked in this auspicious place and their descendants cultivated the memory for almost three millennia.

EXCEPTIONAL LOADS

Two types of material are mainly involved in the transport of megalithic blocks in the Locmariaquer region; an "orthogneiss", a strongly grained and distinctly foliated granitoid rock, which constitutes the region's principal monoliths, including the famous Grand-Menhir, and a "leprous sandstone" almost white, which forms the stone-slab for the apse of the Tables-des-Marchands.
No outcrop of these two rocks is to be found for at least ten kilometres from the peninsular but in both cases it was possible to make at least a part of the journey by raft.

From the Grand-Menhir to the Table-des-Marchands (Table of the Merchants)

Four enormous blocks lying on the ground bear witness to what was perhaps one of the greatest technical achievements of pre-historic man: the extraction, fashioning, transport and erection of a stone monolith more than 20 metres long weighing nearly 280 tonnes. The surface of the giant was carefully studded but the base, destined to be buried, remained rough except for a socket at the end intended to make its positioning easier (Neolithic engineers left nothing to chance!).

The recumbent monster has not failed to rouse the imagination. The breakages separating the now aligned fragments include large splinters undoubtedly due to a deliberate crash. The main breakage, on the other hand, is very abrupt and its sides are even underlined with a

▼ Near the Grand-Menhir another large stele must have stood; its main fragments now cover the chambers of the Table-des-Marchands and Gavrinis.

The entry to the corridor ▸ of the Table-des-Marchands, showing a monumental lintel in the façade of the restored cairn.

The extent of the Er-Grah ▾ tumulus can really only be appreciated from the air. The cavern occupies the centre of a "primary" cairn, extended on each side in a second stage. The whole is clearly positioned at the base of the Grand-Menhir.

series of small crenellations. Lastly the position of the blocks suggest that in throwing it down, the higher part had "chased" the base which, unbalanced, subsequently slowly subsided.

Obviously the "guilty party" has been sought in the attempt to date and explain the fall.

An eighteenth century drawing shows the monolith lying and broken as it is today, while a nautical manual of 1483, drawing attention to the large neighbouring tumuli as landmarks, makes no mention of the menhir, probably already on the ground.

The hunt for pagan symbols at the time of the spread of Christianity or vandalism by Gallo-Romans (Locmariaquer was at the time an important settlement) are possible culprits.

Wind, lightning and earthquakes have been blamed (the Chronicle of Saint-Brieuc in particular relates the serious seismic tremors in the region in 1236).

Archaeological discoveries in recent years have, however, led one to suspect the Neolithic people themselves. A few centuries after the large menhirs were erected in the Locmariaquer region, they might have been subjected to a kind of "iconoclastic fury" before the stones were reused, possibly in some of the dolmens we see today. Excavations have shown that originally the Grand-Menhir did not stand alone: the wedges of others, regularly aligned

along 55 metres seemingly in decreasing size, have been found to the north-east of the giant monolith.

The alignment probably still existed when the Neolithic people undertook the construction of the large **Er-Grah tumulus.** In its present state the latter presents an immense trapeze 140 metres long, 16 metres wide at its northern extremity and 26 at its southern extremity, whereas its height remains low (less than 2 metres). Excavations have shown that this monument was the result of a long and complex history.

From the middle of the fifth millennium BC onwards, several ditches, dwellings and small burial mounds covered the sacrificial burial of the two oxen already mentioned.

Some time afterwards, a large rectangular cairn covered the entirety, protecting a big cavern whose covering slab is still noticeable on the surface of the monument.

Perhaps several decades later, a little before –4000, the cairn was partially elongated, but mainly towards the south in the direction of the Grand-Menhir by heaps of limestone, simply clothed in stones

at the edges and on the surface to match its appearance with the pre-existing monument.

The Table-des-Marchands epresents the third component of this prodigious group. A quarry opened since the Roman era in the 30 x 25 metres oval cairn had revealed the enormous slab covering the funerary chamber, to which the monument owes its name. Excavations in recent years have found sufficient remains to be able to reconstitute the whole; it is necessary, however, to imagine that the cairn with double stone facing was originally at least twice the height of the restoration, with an appearance probably rather close to that of Gavrinis. Rather more recent than Er-Grah, the construction of the Table-des-Marchands could date back to the very beginning of the fourth millennium BC. This time the tomb is not a closed cavern but a "dolmen with corridor", conceived so that it was possible to return to it regularly for new burials or to celebrate the cult of the dead.

When one goes into it, one is struck by the height beneath the roof (nearly 2.5 metres); it is dominated by a large pointed slab in ornamented white sandstone which faces the entry. Excavations have shown that there was a stele previously standing on this spot in the open air (probably it formed part of the system associated with the Grand-Menhir before being incorporated in the new architecture). The chamber's ceiling is formed by a huge slab of orthogneiss (7 metres long, 4 metres wide and 0.8 metres thick), where the visible face is also decorated, but cut off by a breakage. The excavations at Gavrinis, 4 kilometres distant, have

▲ *The breakage of the slab covering the Table-des-Marchands, uncovered during the most recent excavations of the monument.*
Photo: Charles-Tanguy Le Roux.

▲ *On the chamber's ceiling, the representation of an ox, cut by the break in the slab and its splinters, is partially apparent under the top of a pillar.*
Photo: Charles-Tanguy Le Roux.

◄ *The large pointed stele decorated with a crosier design stood in the open in front of the Grand-Menhir alignment before being incorporated in the construction of the Table-des-Marchands.*

▲ *The small broken stele which formerly adorned the Mané-er-Hroeg tumulus provides a summary of Armorican megalithic art: emblem, axes, crosiers and horn shapes.*
Photo: Charles-Tanguy Le Roux.

▲ *The Mané-Lud dolmen contains several ornamented slabs; also the chamber's floor is formed by a large stele with an emblem sinking down into the earth.*

provided the key to the enigma by enabling the reconstitution of another "great-menhir", which had probably been torn from the neighbouring alignment before being cut up to cover the two dolmenic chambers.

A SACRED ART AND ITS SYMBOLS

The sometimes very complex decoration of some of the monolithic megaliths of Morbihan appears to be mainly composed of basic signs; the chief among them must have possessed strong symbolic value:

- the "emblem", more or less complex, which evoked the principal divinity, most likely feminine and the goddess of fertility, life and death;
- the "crozier", probably the simplification of the shepherd's crook was most likely a symbol of authority and domination over the animal world;
- the "axe", displayed in several forms, perhaps already represented the force of destruction (lightning ?) and domination over the vegetable world;
- the "horn formation" must have evoked a bull-like divinity, probably the masculine partner of the goddess of the emblem.

The same "emblem" is enthroned in majesty on the pointed stele of the Table-des-Marchands. It is covered in multiple crosiers, whose meticulous arrangement makes the stone a masterpiece of composition. Another large stele in "emblem" style is lodged facing the ground in the Mané-Lud dolmen, forming the flooring of the sumptuously decorated chamber. However, the tradition of decorated monuments remained lively for a long time in Locmariaquer since a much later monument, the outstanding Pierres-Plates tomb, displays an extraordinary series of signs derived from the same "emblem" but treated in a radically different style, probably dating from the end of the fourth millennium BC.

Fallen giants and giant tumuli

The fury with which the sometimes gigantic standing stones were destroyed a few centuries before being recycled into a second generation of monuments is not the least disconcerting of the characteristics of the major megalithic works at Locmariaquer. Enormous blocks of orthogneiss lie at the feet of the Mané-er-Hroeg or at Bronso, whereas reused steles can be recognised in the dolmens of Mané-Lud, Mané-Rutual and the later monument of Pierres-Plates.

Already in the fifteenth century, the navigator Garcie Ferrande had been struck by the landmarks which stood on the horizon at the entry to the gulf of Morbihan, in a countryside much less wooded and populated than it is today. Like its opposite

The angled tomb of the Pierres-Plates ▶ shows that the Locmariaquer peninsular was still a special cultural sight more than a thousand years after the erection of the first megaliths.

number Arzon, Locmariaquer possesses some large monuments comparable to the Saint-Michel tumulus in Carnac.

At the northern entry to the small town, **the Mané-Lud** is a large, elongated burial ground 80 metres long, the western extremity of which shelters a large dolmen passage way. The monument, however, is mainly famous for the curious alignment of standing stones, topped by the skulls of horses, according to the excavators at the time, which was buried near its eastern extremity.

The Mané-er-Hroeg, at the south of the commune, is more compact in form but, unfortunately, it is now embedded between houses. Its cavern has given up an impressive furnishing in fine polished stones, today in the Vannes museum.

A major centre of megalithic art

Exceptional architecture often calls for exceptional decoration. This was already the case in the Neolithic age; several of the large steles, which have come down to us, were adorned.

The Grand-Menhir itself had the form of an axe, set in relief against the surface's finish.

Complex decoration binding "axes", "a crozier" and images of oxen are displayed on the monolith which was cut up to cover the Tables-des-Marchands and Gavrinis chambers.

A third large stele (11 metres long), which covers the Mané-Rutual dolmen, bears a huge "emblem" whereas at least three others, on a smaller scale, are incorporated in the architecture of the same monument.

The same "emblem" is enthroned in majesty on the pointed stele of the Table-des-Marchands. It is covered in multiple crosiers, whose meticulous arrangement makes the stone a masterpiece of composition. Another large stele in "emblem" style is lodged facing the ground in the Mané-Lud dolmen, forming the flooring of the sumptuously decorated chamber.

However, the tradition of decorated monuments remained lively for a long time in Locmariaquer since a much later monument, the outstanding Pierres-Plates tomb, displays an extraordinary series of signs derived from the same "emblem" but treated in a radically different style, probably dating from the end of the fourth millennium BC.

▼ In the Pierres-Plates, the sign in emblem form has evolved into this strange ornamented buckle, but it probably still symbolises the great Neolithic divinity.
Photo: Charles-Tanguy Le Roux.

In its present state, ▼ the Gavrinis cairn is shown half-emerging from the covering of stones under which it was buried by the Neolithic men.

As long as it has been known, the large cairn, which crowns the southern point of this small island, has always been torn open by an enormous crater at the bottom of which a crevice, it is said, enabled one to worm one's way into a cache of hidden booty. Recent excavations have found pieces of medieval pottery in the opening and it is highly probable that the mound's first explorers were the monks from a small priory located there in the Middle Ages, probably to undertake a vigil facing the entry to the gulf. The real discovery of Gavrinis, however, dates from 1832, when its owner decided to explore the mound at the same time as he installed farmers on the island. More research was then undertaken in 1881 and 1886, then from 1925 to 1927 and finally from 1979 to 1984 into a monument which in 1961 had become the property of the department.

Gavrinis, guardian of the gulf

The islet ▼ of Gavrinis is strategically placed at the entrance of the gulf and alongside the deep channel marking the submerged estuary of the Vannes stream.

From cairn to tumulus

In 1979 the Gavrinis monument had much the same appearance as in the nineteenth century engravings: a mound with a flattened top of about

50 metres in diameter and 7 metres high. On its flank a small door had been set to protect the entry to the dolmen. Research revealed a construction in two major phases.

The heart of the monument comprises a quadrangular cairn with rounded angles, encircled by a facing in dry masonry, strongly inclined towards the interior. The south-east façade, about 30 metres long, is slightly concave, as though to shelter the entry to the dolmen which opens at its centre. Raised in a single flight and still preserved by squares up to nearly 4 metres in height, the monumental façade is sheeted by a series of transversal facings, which structure the shape of the cairn in solid scales on each side of an axial core.

To free the cairn's façade, the excavators had to remove an enormous heap of stones, almost 5 metres thick in places. It was elongated laterally to give the mound its form and exterior dimensions. Sand was also spread on the surface to make its appearance regular. In front of the façade a former flooring had thereby been covered over for nearly 300 m^2. This piling-up had immediately followed the burning of light wooden constructions, standing a short while beforehand on the enclosure. Radiocarbon dating of burnt remains has shown that this probable "condemnation ceremony" had taken place between 3400 and 2900 BC. The monument's explorers, at the beginning of the nineteenth century, had noted that the dolmen's corridor had been packed with loose stones throughout its length which was certainly part of the same process. Unfortunately we know nothing about the contents of the chamber, probably entered from the crater in the Middle Ages.

Unusual dolmenic architecture

The Gavrinis crypt follows the general plan of a dolmen with a long corridor and simple chamber, in the classical style of Armorican megaliths despite some particular features.

The construction displays the desire to make a pure megalith, unlike other monuments, such as the Table-des-Marchands, where masonry was used to finish the structure. Walls, flooring and covering comprised only large stone slabs, obviously chosen to fit exactly and sometimes even cut for the purpose.

From the technical point of view, the achievement is particularly well finished: the uprights had been wedged into a furrow lined with sand, whereas the blocks stand on a filling of loose stones, in some places almost one meter thick, thereby keeping the level regular while improving the structure's stability.

With more than 13 metres, the corridor is longer than others known in Brittany but, curiously, it

GOAT OR ABYSS?

Contrary to what has often been said, Gavrinis is probably not "the isle of the goat" because this name – an ancient one – cannot be directly explained by modern Breton. On the other hand there exists an old Celtic root, Govero, which is associated with the idea of a gully and treacherous waters (ref. the French "gouffre" (abyss)). The island, however, dominates the Gulf of Morbihan at the very place where the impressive eddies of the tidal currents are the most violent and where the channel is particularly deep (about 20 metres). Gavrinis would therefore be the isle of the abyss…

▼ *The Gavrinis monument is remarkable for the disproportion between its various components: tiny chamber, very long corridor, big quadrilateral and enormous circular tumulus.*

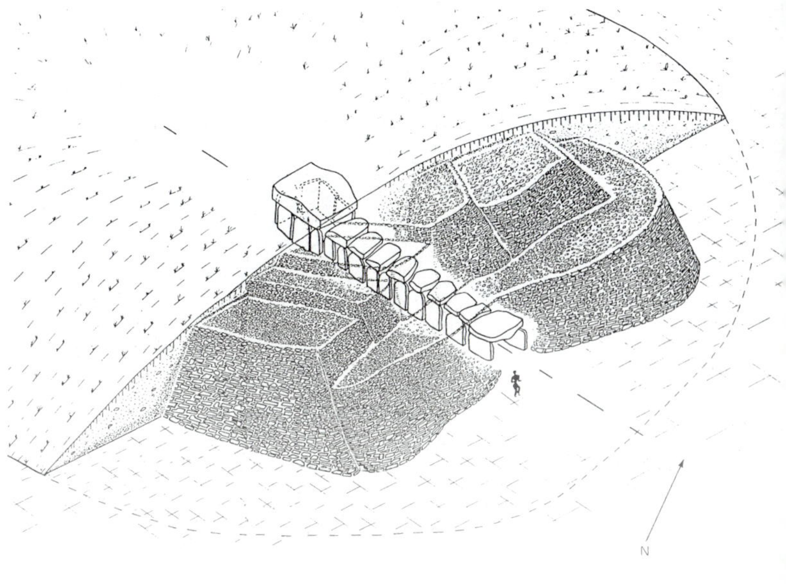

PROSPER MÉRIMÉE TAKES A LOOK

What distinguishes the Gavrinis monument from all the other dolmens I have seen, is that almost all the stones of its walls are sculpted and covered with strange designs. They are curves, straight lines, broken, traced and combined in a hundred different ways... One notices from a few of them that by their regularity and singular positioning they could resemble writing characters... *(Notes from a journey in the west of France, 1835).*

The impression ▲ Mérimée received is still felt by all the visitors to Gavrinis.

Some slabs show ▶ the remains of their first decoration, refashioned to make way for a new composition.

The sign of the axe ▶▶ appears no less than eighteen times on this slab. Some of the blades are closely side by side which probably has a symbolic significance.

leads into a very small chamber (less than 8 m^2).

The corridor opens to the south-east, the most frequent direction for American dolmens but, whatever has been said, rather too much towards the south for the rising sun to be able to penetrate it fully, even at the time of the winter solstice. It is therefore impossible to believe that Gavrinis had been astronomically sited, even more so since the stone slabs and ceiling rise slightly from the entry towards the chamber; at the end of it one can only see the horizon by stooping.

Unrivalled decoration, rich in meaning

More than a century and a half after its discovery, the decoration of Gavrinis remains just as fascinating as it was on the first day, now that the ways of understanding it are beginning to emerge.

Of the 29 stone slabs forming the walls of the crypt, 23 are ornamented, including 21 almost totally. At the entrance, decoration only begins at the third pillar on the left-hand side or the fourth on the right, whereas the "major" panels, owing to the richness of their composition or the quality of their execution, are grouped in the part halfway along the corridor. The decoration of the chamber appears more confused and less well finished in its overall effect. Besides the disconcerting profusion of lines, the key to the decoration lies in the repetition of the basic signs, which we have already described. One can discern the "emblem", the "axe", the "crozier" and the "horn formation".

As at Locmariaquer, the complex compositions combine several

◄ *On the central part of the pillar, an emblem is draped in the form of a "headpiece" which embraces a crosier and an axe blade, probably symbols of the power of the central divinity.*

▼ *1. At the base of the same panel, the axe blades are accompanied by snakes confronting one another.*
2. The cavity in the chamber's slab is a natural fault in the stone, at the base spiral decorations are apparent.
3. At the end of the chamber is another emblem surrounded by crosiers surmounted by two side-by-side axe blades and several fitted-in horn shapes.
4. Halfway along the corridor two side-by-side axes and also a bow and arrow mount symbolic guard against intruders.

1.

2.

3.

4.

signs probably expressing the myths already mentioned. For instance, on the left hand stone slab, at the end of the chamber, an "emblem" seems to sit enthroned in the centre of the upper part between two series of crosiers, above the fitted-in axe-blades and horn shape. As the panels proceed, other more rare signs are recognisable like the large axe, which seems to stand symbolic guard in the corridor, the snakes, which appear on the two uprights, and the enigmatic "pseudo-spirals" on several others. On the other hand one must not fantasize too much about the famous "rings" in the chamber, a simple geological curiosity, but which might perhaps have guided the choice of this particular stone.

Several of the Gavrinis stones were already decorated when they were used to build the monument.

Some of them had been simply recovered as "a stele with crosiers" (on the right of the chamber) which, unlike those in the Table-des-Marchands, was in this case relegated to a lateral wall and even had its top cut off without further alteration, so that it could be included in the construction.

Some of the decorations have been overlaid, as is the case of the stone slabs in the corridor where again the "emblems" and a "crozier" in the ancient style are discernable in the profusion of superimposed ornamentation.

Some stones, such as the right-hand pillar at the end of the chamber have been simply turned round to decorate the other side, the first decoration, relegated to facing the cairn, having only been recovered at the time of the excavations. Sometimes a "declassified" decoration of this kind remains discernible on the edge of a stone slab, like the stripes fitted into the penultimate pillar in the corridor or those in the threshold of the chamber.

▲ *Like the stone at the head of the Table-des-Marchands, this stone must have been a stele with an emblem ornamented with crosiers, but here it has been cut without alteration for incorporation in the construction.*

◄ *On this slab in the corridor a crosier, a large bow and two small emblems probably reveal the first, completely re-done, decoration.*

A large ox with outsize horns ▲ belongs to the decoration of the fragment of the large stele re-used to cover the Gavrinis chamber, now a part of the cairn (ref. p.17).

In the corridor's wall, ▼ a small inserted stone bears an axe with handle, also probably taken from a previous monument.

Lastly we must remember the extraordinary journey made from Locmariaquer by the huge trunk of the orthogneiss stele covering the chamber; its spectacular decoration (two large oxen and a plough-axe), are complimentary to the one in the Table-des-Marchands. It was only discovered buried in the cairn during the 1983 excavations.

Some questions on the site

Likely to occur to the visitor to Gavrinis.

Its construction was carried out with the greatest care but making broad use of components taken from pre-existing monuments – and perhaps carried a long way (the only certainty concerns the main stone, about twenty tonnes, which was taken by sea for at least 4 kilometres; other examples are quite possible, notably in the case of the two quartz pillars).

Its condemnation after only a few centuries, was probably the culmination of an impressive ceremony.

Its monumental nature must have been even more imposing at the time: the solid "tower" of the cairn could have exceeded its present height by one or two metres and the level of the sea, lower than today, would have produced an even more pronounced "low angle" effect.

Lastly, the layout of the crypt, with an abnormally small chamber by comparison with the corridor and wall decoration, is without known equivalent.

Probably the exceptional monument had an exceptional status. It suggests a mausoleum of an outstanding personage, in the tradition of the great tumuli of Carnac dating from the previous millennium. One might also consider the theory that it was an "initiation temple", intended not so much to perpetuate the memory of a dead person but to introduce the living into a sacred circle; the sumptuously ornamented walls still bear witness to abundance and complexity.

By way of conclusion

On a European scale, rare are the privileged regions which can boast a monumental heritage so ancient, impressive and diversified as that of the "coast of the megaliths" at Morbihan, which we have just surveyed. A few exceptional monuments, however, must not make us forget a mass of more modest work. All bear witness to a society of which we are the remote inheritors, the society of the first farmers who dared to clear the extreme west of Europe for cultivation almost seventy centuries ago. The magnificence of some of the tombs still display the power of its elite; the technical skill of some of its achievements still fascinates today's engineers, whereas the architecture and decoration reveal a strong sense of the sacred. It was not only the crises of this society which caused the monuments to be destroyed or transformed. The megaliths remain keystones of our history, let us therefore be sure not to spoil them.

On the edge of this pillar ▶ in the corridor a preliminary striped decoration descends well below the present positioning of the slab.

▼ Owing to the richness of their inspiration and rigour of their composition, some of the Gavrinis stones deserve to be ranked among the masterpieces of the world's art.

Components of a chronological scale
(calibrated dates, in years BC).

Towards -8000:	end of last Ice Age in western Europe, transition from the final Palaeolithic period to the Mesolithic; in the Near East, start of the Neolithic process.
Towards –5500:	arrival of the Neolithic in Alsace, and also in Provence and Languedoc.
Towards –5000:	first indications of the Neolithic process perceptible in Morbihan (Locmariaquer), probably contemporary with the last Mesolithic hunters (Teviec, Hoedic, Quiberon) for a few centuries (perhaps up to –4500?).
Towards –4500:	probable construction of the first megalithic menhirs of Locmariaquer? And tumular burial mounds, (Manio at Carnac, Erdeven), then soon the large tumuli of Carnac (Saint-Michel…) and perhaps the first corridor dolmens (Kercado at Carnac).
Towards –4200:	probable destruction of the Locmariaquer menhirs, at the same time as the construction of the Er-Grah tumulus.
Towards –4000:	zenith of the megalithic period of Morbihan; construction of the majority of the large corridor dolmens (including the Table-des-Marchands) and probable start of the Carnac alignments.
Towards –3800 – 3500:	construction of Gavrinis, and probably the majority of tombs with a corridor and complex chamber (Keriaval type at Carnac and Mané-Groh at Erdeven).
Towards –3400 – 3000:	condemnation of Gavrinis.
Towards –3200 – 2800:	construction of the outstanding tombs (Pierres-Plates type at Locmariaquer) with lateral entry (Kerlescan at Carnac).
Towards –2500 – 2200:	building of the tombs with covered alley (type not represented in the Carnac region where re-use of already existing tombs continued); probable erection of the last menhirs.
Towards –2200 –2000:	probable coexistence between the last users of the megalithic tombs and the first builders of the Bronze Age tumuli (Elven or Priziac type).
Towards –700 – 600:	first Iron Age influences in the region (Rocher necropolis at Bono).
Towards –500 – 400:	main Iron Age development in the region.
Towards –400 – 300:	probable era of the majority of the "Gallic steles" in the region (of the type which adorns the belvedere of the Saint-Michel tumulus).
–56 –50:	Gallic war and Roman conquest.

Practical information

Some information on the main public reception points:

AT CARNAC

• **RECEPTION BUILDING FOR THE KERMARIO ALIGNMENT, 56340 CARNAC,** model of the alignments, bookshop, exhibitions, lecture visits on reservation.
Open everyday throughout the year from10:00 to 17:00 in off season, from 9:00 to 18:00 in high season, from 9:00 to 22:00 in July-August.
Information, reservations, reception:
Tel. 02 97 52 89 99; fax: 02 97 52 60 03

• **MUNICIPAL PRE-HISTORY MUSEUM,** place de la Chapelle, 56340 Carnac (in town centre): regional archaeological collections (from the Palaeolithic to the Middle Age), castings of ornamented megaliths, exhibitions, bookshop.
Open all year:
from June to September, all days from 10:00 to 18:30,
from October to May, from 10:00 to 12:00 and from 14: to 17:00, closed on Tuesday.
Information: Tel.: 02 97 52 22 04
website: www. museedecarnac.com ;
e-mail musee.carnac@wanadoo.fr

AT LOCMARIAQUER

Reception centre for the Tables-des-Marchands, Kerlogonan road, 56740 Locmariaquer (access from the town or the west byroad): audiovisual, exhibitions, bookshop, guided visit.
Open throughout the year except from 20 December to 10 January approximately; hours vary according to season.
Information, reservations:
Tel. 02 97 57 37 59; fax 02 97 57 41 62.

FOR GAVRINIS

Visits with restricted numbers with reservation at Larmor-Baden from April to October. Reception: cale de Pen-Lannic, 56780.
Information, reservations:
Tel./fax 02 97 57 19 38.

AT VANNES

Pre-history museum at the Château-Gaillard, 2 rue Noé, 56000 Vannes (Morbihan archaeological collections, in the process of assignment to the museum centre of the town of Vannes, visiting hours being arranged).
Information:
Tel. 02 97 47 35 86; fax 02 97 47 00 39.

Bibliography

Gérard BAILLOUD, Christine BOUJOT, Serge CASSEN and Charles-Tanguy LE ROUX: Carnac. *Les Premières Architectures de pierre.* CNRS Editions, Paris, 126 p. (new edition in preparation).

Jacques BRIARD: *Les Mégalithes de l'Europe atlantique.* Errance, Paris, 1995, 128 p.

Pierre-Roland GIOT: *Bretagne des mégalithes.* Editions Ouest-France, Rennes, 1995, 128 p.

Pierre-Roland GIOT, Jean-Laurent MONNIER and Jean LHELGOUACH: *Préhistoire de la Bretagne.* Editions Ouest-France, Rennes, 590 p. (2nd edition, 1998)

Charles-Tanguy LE ROUX: *Gavrinis.* Gisserot, Paris, 1995, 32 p. Jean LHELGOUAC'H *Locmariaquer.* Gisserot, Paris, 32 p.

Jean-MOHEN *Les Alignements de Carnac.* Editions du Patrimoine, Paris, 2000, 64 p.

Anne-Elisabeth RISKINE *Carnac. L'armée de pierres.* Imprimerie nationale, Paris, 118 p.

Anne-Elisabeth RISKINE: *Le Musée de préhistoire de Carnac.* Editions Ouest-France, Rennes, 1987, 32 p.

Cet ouvrage a été achevé d'imprimer par T.P.I. (Technic Plus Impression) à Betton (35)
Maquette : B. Racine - Cartographie : Patrick Mérienne
ISBN 2.7373.2859.4 - N° d'éditeur : 4234.01.03.03.01 - Dépôt légal : mars 2001

Only the monuments mentioned in the text appear on this map